No Metal No Magic
Element 3

Lithium, Presented by Lillian

From The Magical Elements of the Periodic Table Series

By Sybrina Durant with Illustrations by Pranavva et al.

Lithium, Presented By Lillian
From The Magical Elements of the Periodic Table Series

Story copyright 2025

Soft Cover Print ISBN- 978-1-942740-53-7

BISAC Codes:

JNF051070 JUVENILE NONFICTION / Science & Nature / Chemistry

JNF016000 JUVENILE NONFICTION / Curiosities & Wonders

JNF051080 JUVENILE NONFICTION / Science & Nature / Earth Sciences / General

Soft Cover Print ISBN 13 - ISBN: 978-1-942740-53-7

All rights reserved by Sybrina Publishing and Distribution Company.

League City Texas, United States of America

This book contains material protected under International and Federal Copyright Laws and Treaties. Any unauthorized reprint use of this material is prohibited.

No part of this book may be reproduced or transmitted in any form or by any means, electronic or mechanical, including photocopying, recording, or by any information storage and retrieval system without written permission from Sybrina Publishing and Distribution Company.

Contact Sybrina@sybrina.com.

Lillian The Elemental Dragon Presents Lithium

This Element 3 book features the periodic table element, Lithium. It is presented by Lillian, a member of the Elemental Dragon Clan. Each dragon has a magical tail tipped with an element that gives them unique powers. Their powers are based on the properties of its periodic table element.

Lillian is just one of the 118 elementals who will present all of the Magical Elements of the Periodic Table to readers who are curious about the wonders of the world.

Lillian introduces Lithium in her book.

The Elemental Dragon Clan and their other techno-magical friends are the perfect group to introduce you to the elements in the Periodic Table. Hopefully, this Magical Elements of the Periodic Table book will spark an interest in the magical and real world properties of all the elements known today. You may be surprised at how prominently they feature in our every day lives.

Each page in this book contains terms that might not be completely familiar to the reader. Refer to the definitions in the back of the book to get a clear understanding of each meaning.

There is also a fun elemental themed Periodic Table at the back of the book. It features 118 elements presented by fanciful characters like unicorns, dragons, wizards, knights and goblins.. They want you to remember that if there's no metal...there's no magic or technology.

Remember, "No metal – No Magic. . .and No Technology".

It's Techo-Magical.

Note: Sybrina Publishing websites are Sybrina.com and MagicalPTElements.com. Follow sybrinapublishing on Instagram, Magical Elements of the Periodic Table on Facebook, @sybrinad on Pinterest, Sybrina_SPT on Twitter; and Sybrina Durant on LinkedIn.

Lillian

The Dragon With The Lithium Tipped Tail

Symbol: Li   Atomic Number: 3   Atomic Mass: 6.94

Lithium resides in Group 1 Period 2 on the Periodic Table.

The atomic symbol is Li. It's Atomic Number is 3. It's Atomic Mass is 6.94.

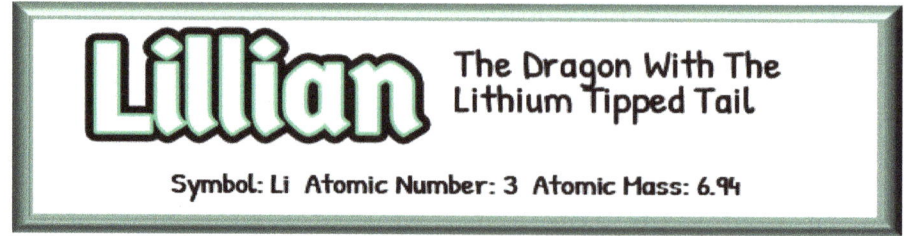

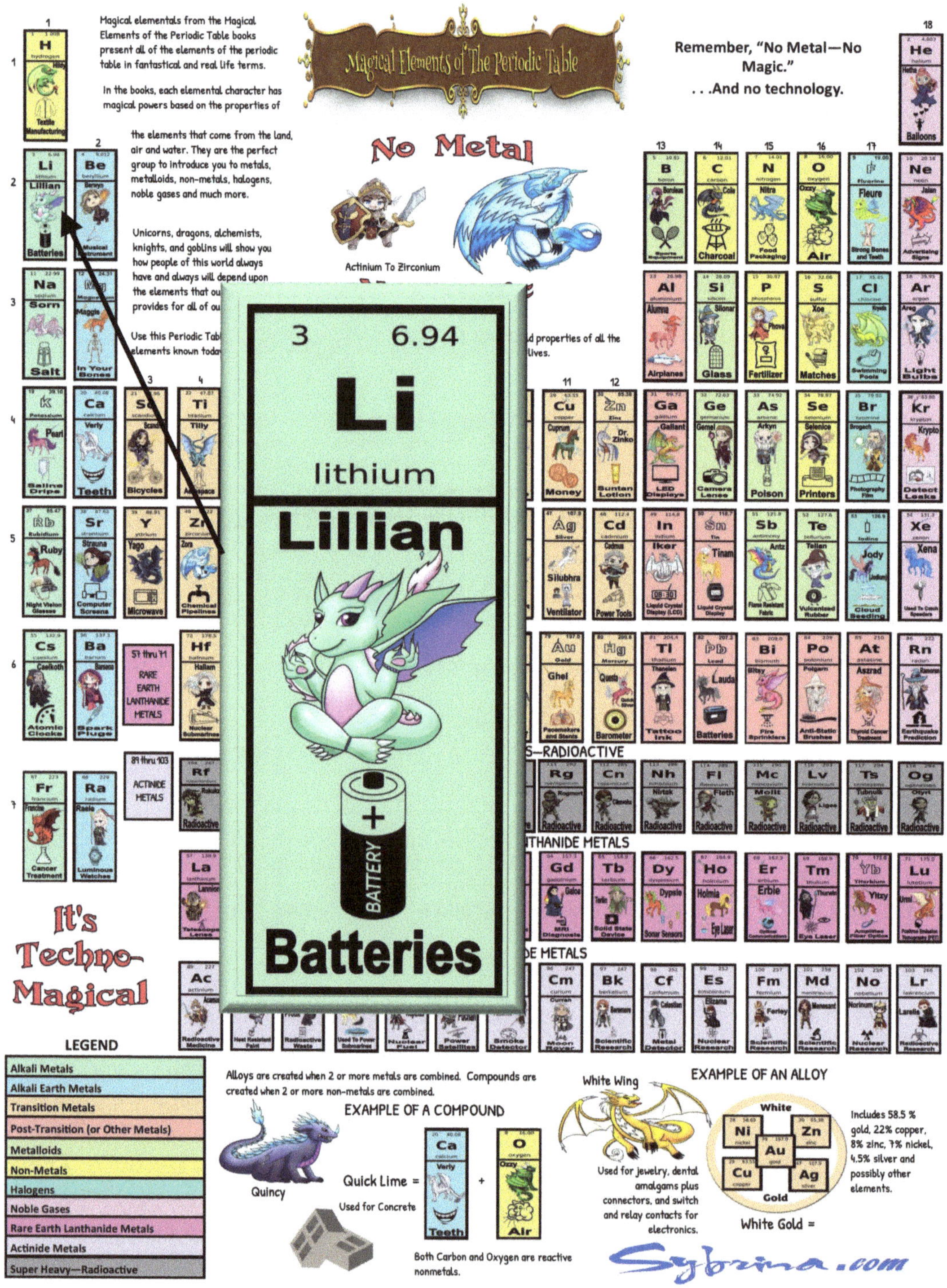

Lithium is an Alkali Metal

Lithium was discovered in 1817 by Swedish chemist Johan August Arfvedson while studying the mineral petalite on Utö island (some say Stockholm), Sweden. He later found lithium in spodumene and lepidolite.

Lithium is a soft, silvery-white, metal that reacts violently with water. It must be stored in air tight environments.

Lithium is a good conductor of heat and electricity. Heat can flow through it pretty quickly. Free movement of electrons enables lithium to conduct electricity effectively.

Lithium is paramagnetic, meaning it has a slight magnetic response but isn't as magnetic as iron or nickel. Lithium ions are non-magnetic and exhibit diamagnetism.

Lithium is a ductile metal, meaning it can be stretched into wires without breaking. It is also malleable, allowing it to be hammered into thin sheets.

Lithium is an Alkali Metal. It is the lightest of all metals.

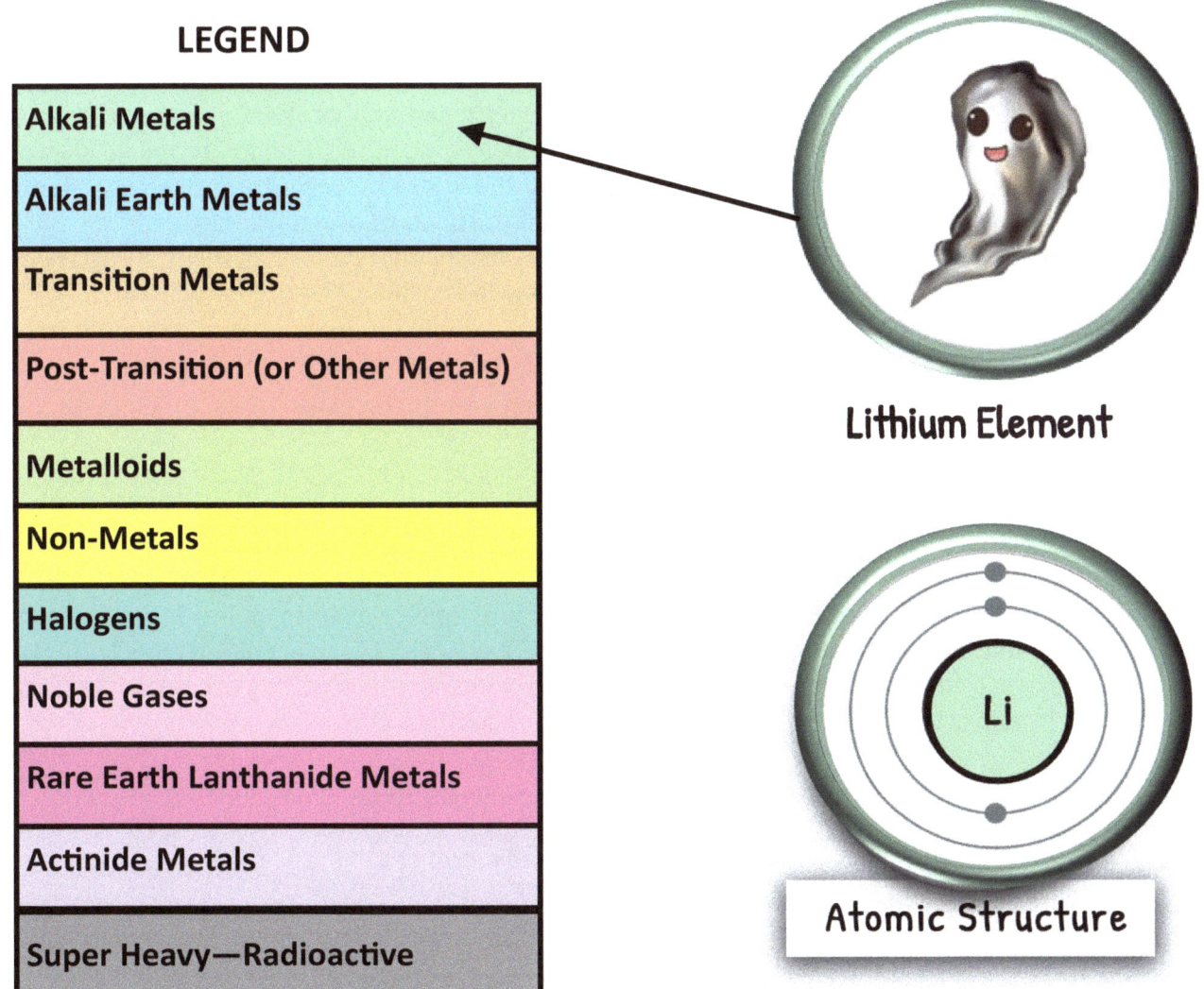

Alkali Metals—Some metals on the periodic table are soft and shiny. They are so soft that they can be cut with a knife! These metals are excited to give away electrons to elements in need, making them highly reactive. This electron transfer creates a compound known as a salt. Surprisingly, these metals are not found in nature alone; they must be extracted from other sources. Examples of these metals include lithium, sodium, potassium, rubidium, cesium, and francium.

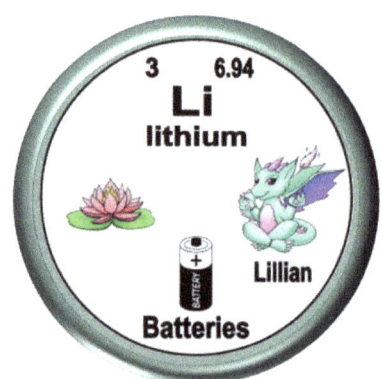

Lithium is a crucial element in our modern world, playing a key role in many technologies we use every day. Known as the lightest metal, lithium's properties make it an exceptional energy storage solution. Devices like smartphones, laptops, and electric vehicles (EVs) all depend on lithium-ion batteries, celebrated for holding considerable energy while being lightweight. This characteristic enables the design of sleek, compact devices that can last longer without needing a recharge, catering to our fast-paced, mobile lifestyles. The efficiency of lithium-ion technology has transformed our energy consumption habits, providing lasting power that fits our on-the-go demands.

Historically, the uses of lithium began to be recognized in the 19th century, primarily in the field of medicine. The earliest known application of lithium as a therapeutic agent dates back to the 1810s when it was first used to treat gout and various nervous disorders. By the mid-19th century, lithium carbonate had gained popularity as a cure-all tonic, marketed to alleviate a range of ailments, making it a groundbreaking substance in early therapeutic practices. This medicinal application paved the way for further exploration into lithium's potential, eventually leading to its role as a mood stabilizer in the treatment of bipolar disorder in the late 20th century. This evolution illustrates how lithium shifted from a curative element steeped in anecdotal claims to a scientifically supported medication, underscoring its importance in mental health treatment today.

In addition to its medicinal uses, lithium's initial applications in various industrial processes have also had profound impacts on technology and manufacturing. The introduction of lithium into glass and ceramics significantly improved production techniques; it lowers the melting point of the raw materials, which increases the efficiency of production and elevates the quality of the final products. Lithium is also utilized in lubricating greases that endure extreme temperatures and pressures, owing to its ability to maintain consistency and performance where traditional oils would fail. These early industrial applications set the stage for lithium's later breakthrough as a cornerstone of modern energy storage solutions, marking it as a crucial element in the evolution of industrial chemistry.

As we glance toward the future, the possibilities for lithium usage continue to expand and diversify. Current research is examining its potential in powering emerging technologies, such as solid-state batteries, which promise enhanced energy density and safety compared to traditional lithium-ion batteries. Solid-state batteries replace the liquid electrolyte used in conventional batteries with a solid electrolyte, potentially reducing the risk of flammability and allowing for faster charging times. This innovation could lead to longer-lasting batteries suitable for electric vehicles, thereby further accelerating the transition from fossil fuels to renewable energy sources. Such advancements will likely play a vital role in reducing global dependence on fossil fuels while simultaneously promoting cleaner energy consumption.

Moreover, as the world grapples with the pressing challenges posed by climate change, lithium's role in renewable energy storage solutions will become even more critical. The integration of solar and wind energy into our energy grid hinges significantly on efficient storage systems that can store excess electricity generated during peak production periods and release it during times of high demand. With ongoing investments in grid-scale battery technology and advancements in lithium extraction methods that are more sustainable and environmentally friendly, the future could see lithium becoming an even more integral player in the renewable energy landscape. The convergence of these factors positions lithium not only as a vital mineral for today's technologies but as a cornerstone of tomorrow's sustainable energy solutions.

The demand for lithium continues to rise in advanced applications such as artificial intelligence, aerospace, and the burgeoning field of quantum computing. In aerospace, lithium's lightweight properties are invaluable, as they can significantly reduce aircraft and spacecraft weight, enhancing fuel efficiency and performance. As industries seek innovative ways to innovate and optimize, lithium's capacity to improve energy efficiency and support high-performance systems places it at the forefront of technological advancements. Likewise, the intersection of lithium with quantum technology could unlock groundbreaking applications in computing and data storage, paving the way for next-generation devices that operate with unprecedented speed and efficiency.

Lithium's trajectory from a historical medicinal compound to a central element in modern energy technology marks a significant evolution in its importance. Its multifaceted applications across diverse industries underscore how lithium will continue to be pivotal in shaping not only our current technological landscape but also guiding us toward a more sustainable future. With ongoing advancements in science and engineering, the significance of lithium is poised to grow exponentially, responding to and addressing the pressing challenges of our time while unlocking new possibilities for technological innovation. As such, lithium will undoubtedly remain a driving force in our quest for ecological sustainability and a cleaner, greener future.

Uses For Lithium

Small rechargeable Lithium batteries are used in pacemakers, digital cameras, smart phones, laptops, watches and more. Large lithium batteries are used in scooters, golf carts, trolleys, boats, cars, trucks, recreational vehicles and more. Small rechargeable lithium batteries play a crucial role in the functionality of various everyday devices, enabling convenience and portability in our fast-paced lives. Meanwhile, large lithium batteries power a range of vehicles and equipment, driving innovation in electric transportation and contributing to the growing shift towards sustainable energy solutions.

Lithium is a medication that effectively reduces manic symptoms and is a treatment for bipolar episodes. It works by stabilizing mood fluctuations, making it an essential component of long-term management for individuals with bipolar disorder. Additionally, regular monitoring of lithium levels is crucial, as it requires careful dosage adjustments to ensure its efficacy while minimizing potential side effects.

Uses For Lithium

(Continued)

Lithium is used to insulate windows from the sun due to its unique properties that reflect and absorb heat effectively. This not only helps in maintaining a comfortable indoor temperature but also reduces energy costs for cooling, making buildings more energy-efficient.

Lithium is an essential ingredient for some specialty glasses and glass-ceramics like glass cook-top stoves. It is responsible for the glass-ceramic's low or near zero thermal expansion, enabling their use in high-temperature ranges without voltage breakage.

The Source of Lithium

Lithium is extracted from Spodumene and other ores like petalite and lepidolite. It's also found in great quantities in salt water. Australia and Chile are the world's largest producers of lithium.

Lithium is a fascinating element that plays a critical role in our daily lives, particularly in the technology and healthcare sectors. This highly reactive alkali metal was first discovered in 1817 by the Swedish chemist Johan August Arfwedson. While he was examining a mineral called petalite, which is rich in lithium, he found this new element. It was a significant moment in the scientific community, as it sparked interest and exploration into lithium's properties and potential uses. Today, lithium is integral to various industries, especially for batteries, which power our smartphones, laptops, and electric vehicles, as well as its use in some medications.

Lithium is found naturally in different environments. It mainly occurs in mineral deposits and brine sources. The minerals that contain lithium include spodumene, lepidolite, and petalite. These minerals are often located in pegmatite formations, which are a type of igneous rock. Countries like Australia, China, and Zimbabwe are significant producers of lithium-bearing minerals, meaning they extract these valuable materials from the Earth's crust.

Additionally, lithium-rich brines can be found in salt flats, also known as salars. These brines are saline solutions that contain a high concentration of lithium and can be found in areas like South America. Argentina, Bolivia, and Chile form a well-known region called the "Lithium Triangle." This geographic area is famous for its vast lithium reserves and other essential minerals, making it a focal point for lithium production.

The Source of Lithium (continued)

When it comes to commercial extraction, there are two primary methods for obtaining lithium: hard rock mining and brine extraction. Each method has its unique processes and characteristics.

In hard rock mining, spodumene is the mineral that most commonly yields lithium. The extraction begins with the mining of this mineral, which involves traditional mining techniques to obtain the ore from underground. Once the spodumene ore is collected, it is crushed into smaller pieces to prepare it for the next steps. Following this, the crushed ore is heated in a furnace at a high temperature of around 1,000 degrees Celsius. This process is known as calcination and serves to convert the lithium into a form that is easier to extract.

After calcination, the next step involves leaching, which is where solvents come into play. Sulfuric acid is often used during this phase to dissolve the lithium from the ore. The resulting solution contains lithium in a soluble form. Following this, the lithium must be purified to remove any impurities, ensuring that it meets the high standards needed for various applications. Eventually, through a series of chemical reactions and processes, lithium carbonate or lithium hydroxide is isolated, depending on its intended use. Lithium carbonate is typically used in ceramics and glass, while lithium hydroxide is crucial for producing batteries.

On the other hand, the extraction of lithium from brine is generally considered less energy-intensive and more environmentally friendly than hard rock mining. The process begins by pumping lithium-rich brine from the underground aquifers to the surface. Once it reaches the surface, the brine is spread out in large evaporation ponds. Here, the sun and wind work together to evaporate water from the brine over several months or even years. As the water evaporates, the concentration of lithium increases, along with other salts that precipitate out of the solution.

Once the brine reaches a desired concentration of lithium, further processing is required to isolate lithium carbonate or lithium hydroxide. This step is similar to that in hard rock extraction, where processing techniques are employed to ensure purity and quality.

While both extraction methods have proven effective in sourcing lithium, they do come with environmental considerations. Hard rock mining can lead to habitat destruction and pollution due to the heavy machinery used and the chemical processes involved. Likewise, extracting lithium from brine can cause water depletion, which affects local ecosystems and communities that rely on these water resources. Therefore, as demand for lithium continues to grow—especially with the rise in electric vehicles and renewable energy storage—addressing these environmental concerns is paramount. Innovations in mining practices and sustainable technologies will play a vital role in ensuring that lithium extraction meets the world's needs while minimizing its impact on the planet.

In conclusion, lithium extraction is a vital process with significant implications for the future. From its discovery by Arfwedson to the rapid technological advancements we see today, lithium has shaped many aspects of modern life. Understanding the processes through which this element is mined and manufactured not only highlights its importance but also emphasizes the need for responsible sourcing as we move forward.

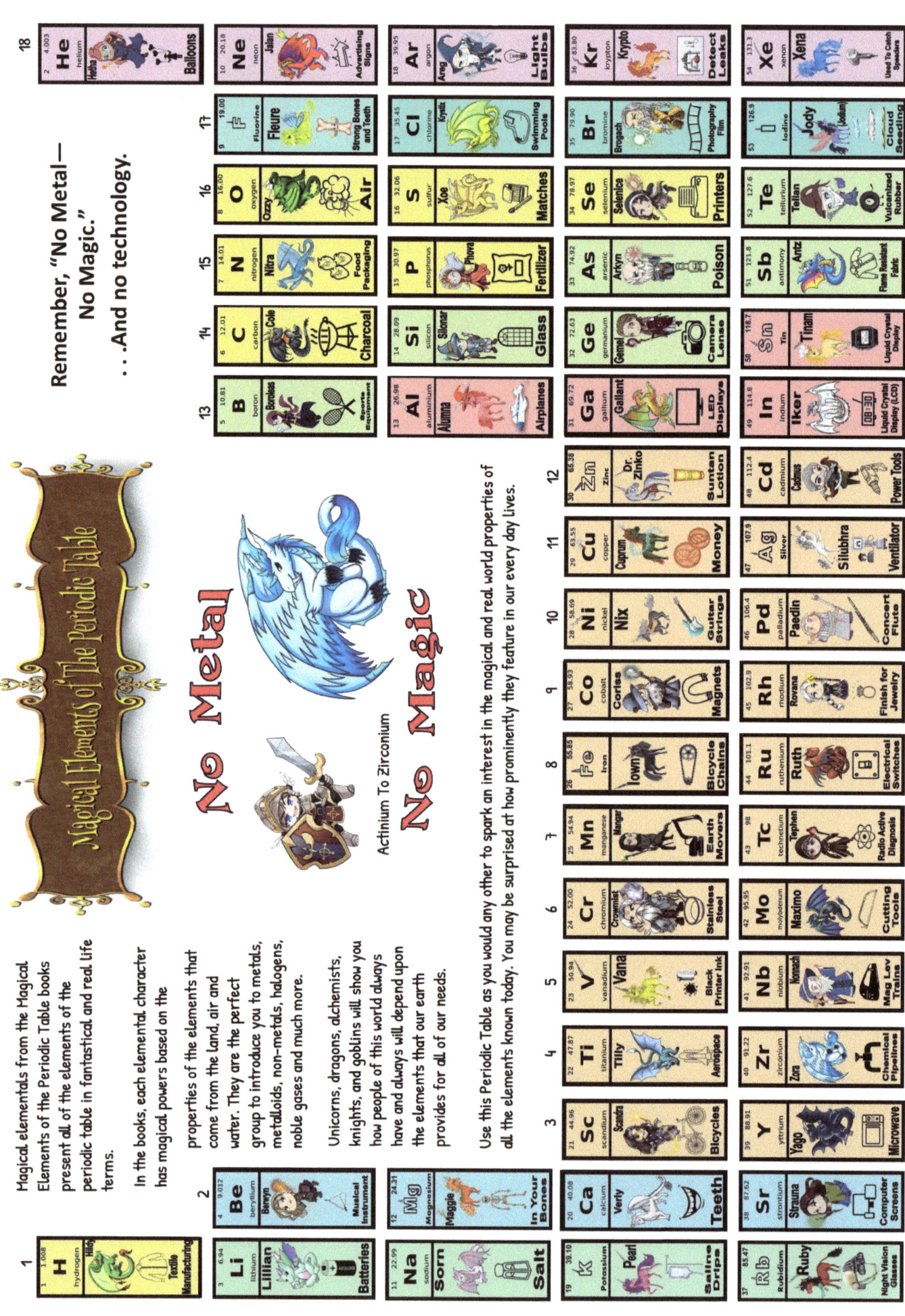

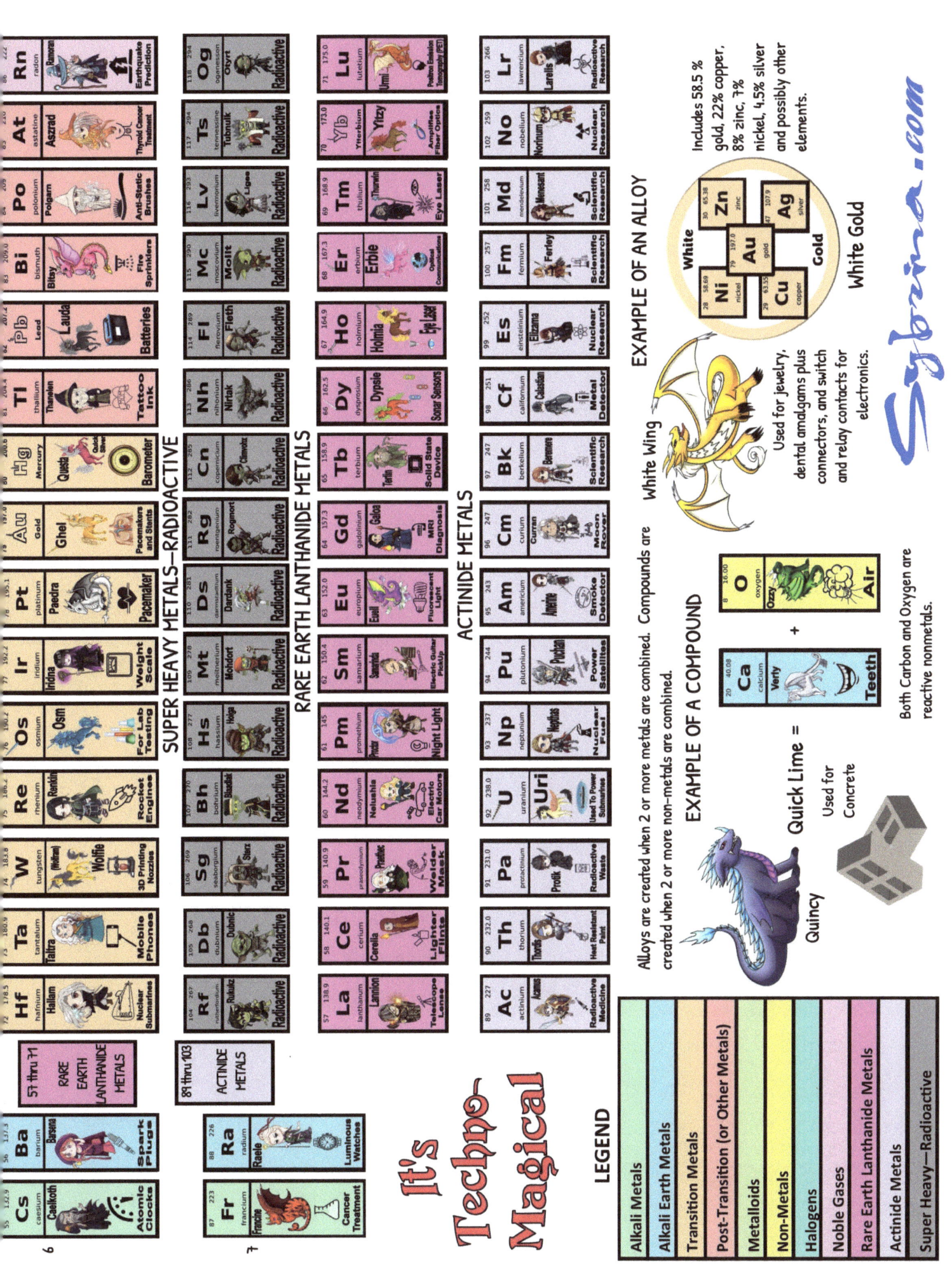

All Of The Periodic Table Elements Listed Alphabetically
Element Listed In Red Is Featured In This Book

ACTINIUM—AC—89

ALUMINUM—AL—13

AMERICIUM—AM—95

ANTIMONY—SB—51

ARGON—AR—18

ARSENIC—AS—33

ASTATINE—AT—85

BARIUM—BA—56

BERKELIUM—BK—97

BERYLLIUM—BE—4

BISMUTH—BI—83

BOHRIUM—BH—107

BORON—B—5

BROMINE—BR—35

CADMIUM—CD—48

CALCIUM (Vital)—CA—20

CALIFORNIUM—CF—98

CARBON—C—6

CERIUM—CE—58

CESIUM—CS—55

CHLORINE (Keen)—CL—17

CHROMIUM—CR—24

COBALT—CO—27

COPERNICIUM—CN—112

COPPER—CU—29

CURIUM—CM—96

DARMSTADTIUM—DS—110

DUBNIUM—DB—105

DYSPROSIUM—DY—66

ERBIUM—ER—68

EINSTEINIUM—ES—99

EUROPIUM—EU—63

FERMIUM—FM—100

FLEROVIUM—FL—114

FLUORINE—F—9

FRANCIUM—FR—87

GADOLINIUM—GD—64

GALLIUM—GA—31

GERMANIUM—GE—32

GOLD—AU—79

HAFNIUM—HF—72

HASSIUM—HS—108

HELIUM—HE—2

HOLMIUM—HO—67

HYDROGEN—H—1

INDIUM—IN—49

IODINE (JODIUM)—I—53

IRIDIUM—IR—77

IRON—FE—26

KRYPTON—KR—36

LANTHANUM—LA—57

LAWRENCIUM—LR—103

LEAD—PB—82

LITHIUM—*LI*—3

LIVERMORIUM—LV—116

LUTETIUM (Unique)—LU—71

MAGNESIUM—MG—12

MANGANESE—MN—25

MEITNERIUM—MT—109

MENDELEVIUM—MD—101

MERCURY (QUICK SILVER)—HG—80

MOLYBDENUM—MO—42

MOSCOVIUM—MC—115

NEODYMIUM—ND—60

NEON (Jazzy)—NE—10

NEPTUNIUM—NP—93

NICKEL—NI—28

NIHONIUM—NH—113

NIOBIUM—NB—41

NITROGEN—N—7

NOBELIUM—NO—102

OGANESSON—OG—118

OSMIUM—OS—76

OXYGEN—O—8

PALLADIUM—PD—46

PHOSPHORUS—P—15

PLATINUM—PT—78

PLUTONIUM—PU—94

POLONIUM—PO—84

POTASSIUM—K—19

PRASEODYMIUM—PR—59

PROMETHIUM—PM—61

PROTACTINIUM—PA—91

RADIUM—RA—88

RADON—RN—86

RHENIUM—RE—75

RHODIUM—RH—45

ROENTGENIUM—RG—111

RUBIDIUM—RB—37

RUTHENIUM—RU—44

RUTHERFORDIUM—RF—104

SAMARIUM—SM—62

SCANDIUM—SC—21

SEABORGIUM—SG—106

SELENIUM—SE—34

SILICON—SI—14

SILVER—AG—47

SODIUM—NA—11

STRONTIUM—SR—38

SULFUR (Xanthous)—S—16

TANTALUM—TA—73

TECHNETIUM—TC—43

TELLURIUM—TE—52

TENNESSINE—TS—117

TERBIUM—TB—65

THALLIUM—TI—81

THORIUM—TH—90

THULIUM—TM—69

TIN—SN—50

TITANIUM—TI—22

TUNGSTEN—W (WOLFRAM)—74

URANIUM—U—92

VANADIUM—V—23

XENON—XE—54

YTTERBIUM—YB—70

YTTRIUM—Y—39

ZINC—ZN—30

ZIRCONIUM—ZR—40

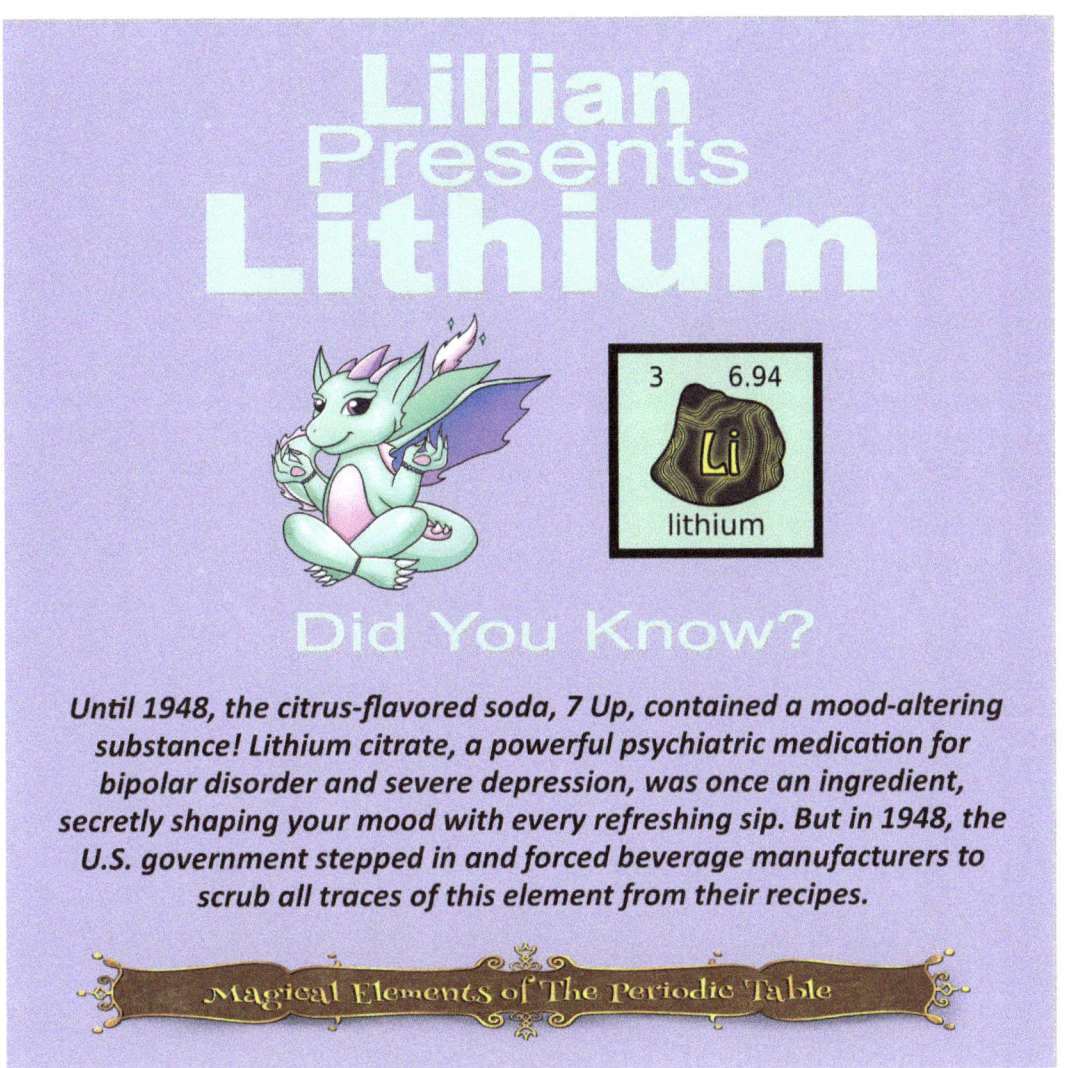

- Lithium is named from the Greek word "lithos," which means "stone," reflecting its discovery in mineral form rather than in plant or animal matter. Its naming emphasizes its association with minerals, highlighting its status as the lightest metal and the first alkali metal in the periodic table.
- Pure lithium metal is extremely corrosive and requires special handling. Because it reacts with air and water, the metal is stored under oil or enclosed in an inert atmosphere. When lithium catches fire, the reaction with oxygen makes it difficult to extinguish the flames.
- Lithium, a lightweight and highly reactive metal, is utilized in the production of foldable glass due to its ability to improve the flexibility and durability of the material. When incorporated into the glass matrix, lithium helps to enhance thermal resistance and reduce brittleness, allowing for greater bending without breakage. This innovative application of lithium not only advances the technology of flexible screens but also contributes to the broader development of foldable electronic devices.
- An Aluminum Lithium alloy's lightweight and strong properties make it an ideal material for the construction of military aircraft, particularly in facilitating fuel efficiency and enhancing maneuverability. As researchers seek ways to improve aircraft performance, lithium alloys are incorporated into fuselage skin and wings to reduce weight without compromising structural integrity. This innovation contributes significantly to the advancement of military aviation technology, allowing for faster and more agile planes

Structure Of Elements In The Periodic Table

Periodic tables are laid out in rows and columns.

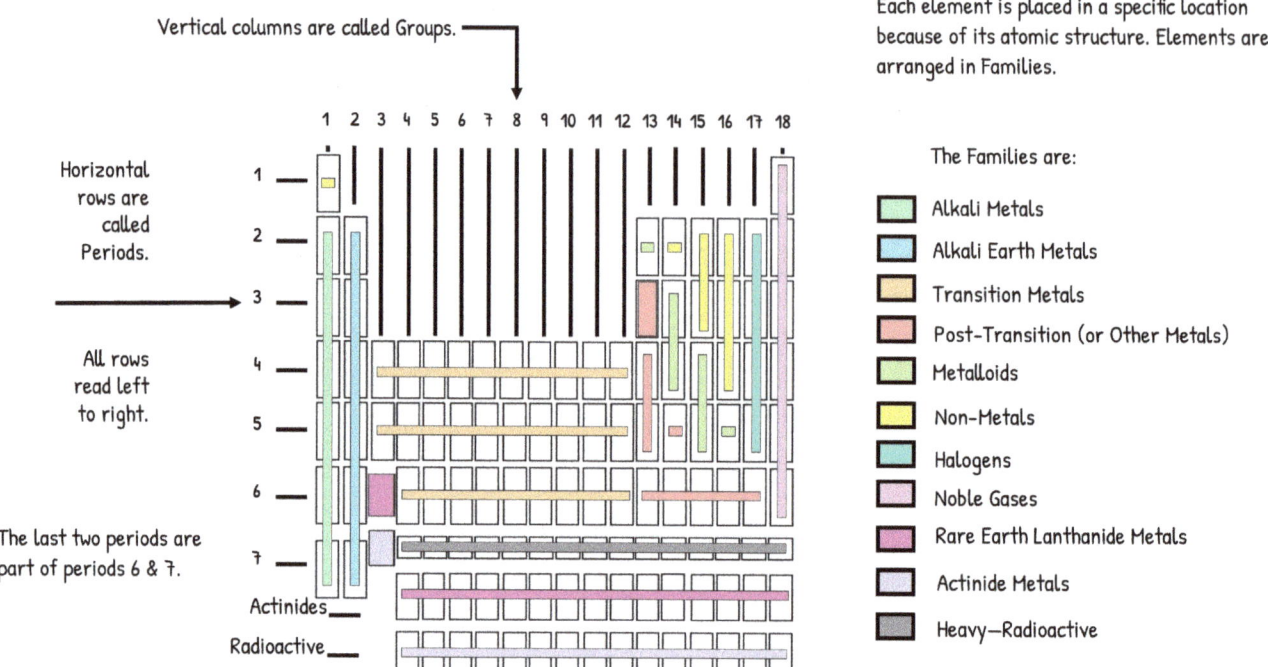

The term 'Element' is used to describe atoms with specific characteristics.
Every element in the first column or Group has 1 electron in the outer orbital (shell).
Every element in the second column (group two) has two electrons in the element's outer orbital.
The number designation of each Group represents the number of electrons in the element's outer orbital—
except for Group 18, Period 1—Helium. It only has 2 electrons.
Those electrons, called Valence Electrons, are what chemically bond with other elements.

Atomic Structure of Element: The atomic structure of an element refers to the arrangement of protons and neutrons in the nucleus of the atom, and the electrons in the electron cloud around the nucleus. Group 1, Period 1—Hydrogen is the only element that has no neutrons.

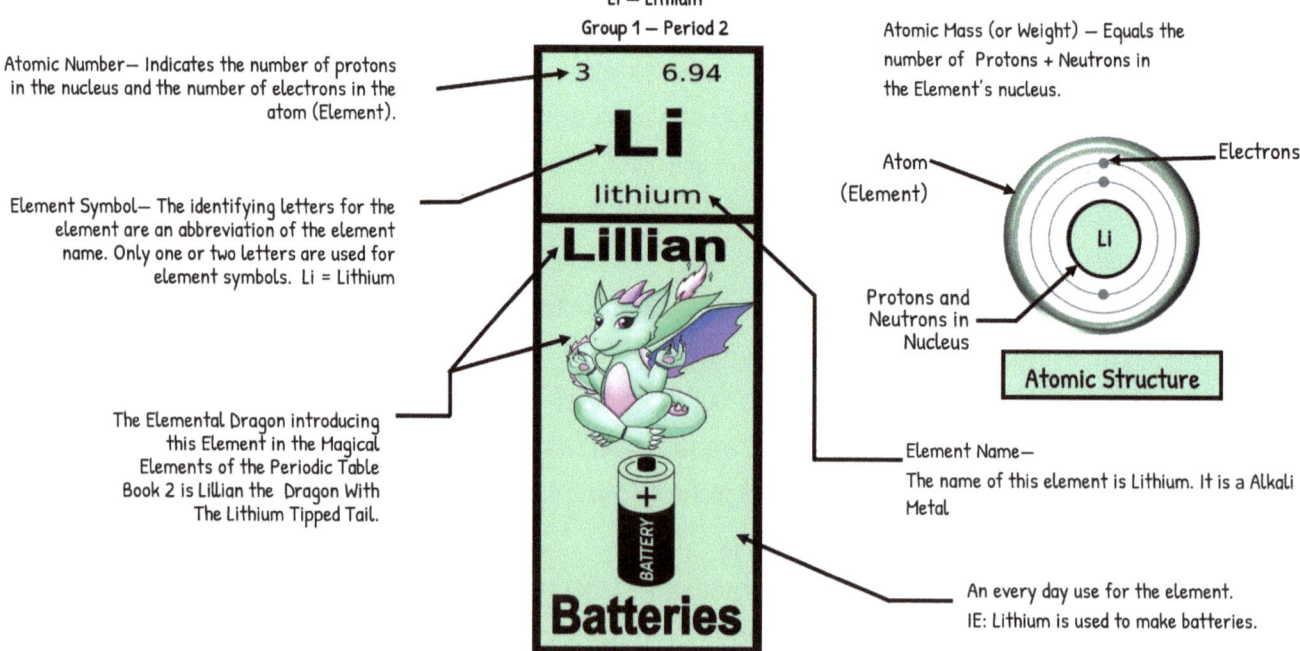

Types of Elements On The Periodic Table

Alkali Metals—Some metals on the periodic table are soft and shiny. They are so soft that they can be cut with a knife! These metals are excited to give away electrons to elements in need, making them highly reactive. Ther electron transfer creates a compound known as a salt. Surprisingly, these metals are not found in nature alone; they must be extracted from other sources. Examples of these metals include lithium, sodium, potassium, rubidium, cesium, and francium.

Alkali Earth Metals—The elements in column 2 of the periodic table have 2 outer electrons in their shell. Ther makes them very active with nonmetals that need electrons to stay stable. When they react, they make something called a salt. They are often found in nature all by themselves, and they can even conduct electricity. The elements are beryllium, magnesium, calcium, strontium, barium, and radium.

Post-Transition (or other Metals)— Elements directly to the right of the transition metals. They are known as "poor metals: and are soft and brittle. These include aluminum, gallium, indium, tin, thallium, lead, bismuth, zinc, cadmium and mercury.

Transition Metal—The main metals are found in the middle and bottom rows of the periodic table. They look like metal, can conduct electricity, can bend and be shaped easily. The period 4 transition metals are scandium, titanium, vanadium, chromium, manganese, iron, cobalt, nickel, copper, and zinc. The period 5 transition metals are yttrium, zirconium, niobium, molybdenum, technetium, ruthenium, rhodium, palladium, silver, and cadmium. The period 6 transition metals are lanthanum, hafnium, tantalum, tungsten, rhenium, osmium, iridium, platinum, gold, and mercury. The period 7 transition metals are the naturally-occurring actinium, and the artificially produced elements rutherfordium, dubnium, seaborgium, bohrium, hassium, meitnerium, darmstadtium, and roentgenium.

Metalloids—The elements called metalloids are a mix of metals and nonmetals. They look like metals, but can't conduct electricity very well. They also break easily and act like nonmetals. These include boron, silicon, germanium, arsenic, antimony, tellurium, astatine, and polonium.

Non-Metals—These elements reside in columns 15-17, and can be gases, liquids, or solids. They don't conduct heat or electricity. The solids are brittle, and they have no metallic luster. They readily accept electrons from metals to form salts. These include nitrogen, oxygen, fluorine, chlorine, bromine, and iodine.

Halogens—Halogen chemicals are a special type of element. When they mix with metal, they become a kind of salt. Halogens are super reactive because they like to take an electron from metals. They can be found in column 17 of the element table. Some of them can be found in nature, but most are very dangerous and can hurt you if you touch them. They include fluorine, chlorine, bromine, iodine, and the radioactive elements astatine and tennessine.

Noble Gases—These elements reside in column 8. They are all odorless, colorless gases that are chemically very stable (inert). They don't generally form compounds by bonding with another element. These include helium, neon, argon, krypton, xenon, and radon.

Lanthanide Rare Earth Minerals—The Japanese call them "the seeds of technology." The US Department of Energy calls them "technology metals." These elements have atomic numbers 57-71. They are vital to industry. They can be added to metals to strengthen them to make alloys such as stainless steel, used to refine crude oil, and are crucial in producing technology—electronics, telecommunications, and metal devices to name a few. They are lanthanum, cerium, praseodymium, neodymium, promethium, samarium, europium, gadolinium, terbium, dysprosium, holmium, erbium, thulium,

Actinide Metals—Any of a series of chemically similar metallic elements with atomic numbers ranging from 89 (actinium) to 103 (lawrencium). All of these elements are radioactive, and two of the elements, uranium and plutonium, are used to generate nuclear energy. The lanthanides and actinides are sometimes called the inner transition metals, referring to their properties and position on the table. They are actinium, thorium, protactinium, uranium, neptunium, plutonium,

Super Heavy—Radioactive—Superheavy elements are those elements with a large number of protons in their nucleus. Elements with more than 92 protons are unstable; they decay to lighter nuclei with a characteristic half-life. They do not occur in large quantities (if at all) naturally on earth, and only exist briefly under highly controlled circumstances. They include lawrencium, rutherfordium, dubnium, seaborgium, bohrium, hassium, meitnerium, darmstadtium, roentgenium, copernicium, nihonium, flerovium, moscovium, livermorium, tennessine, and oganesson.

Polyatomic Ions

While individual elements are typically not polyatomic, certain elements can form polyatomic molecules or ions. Many polyatomic ions exist, formed by groups of atoms covalently bonded together with an overall charge. Polyatomic ions carry a net electric charge, either positive (cation +) or negative (anion -). Despite being made of multiple atoms, polyatomic ions behave as a single, distinct entity in chemical reactions and compounds.

Ammonium =
(NH_4^+)
Contains one nitrogen and four hydrogen atoms.

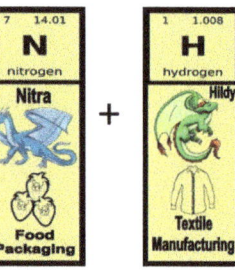

Ammonium is widely used in agriculture as a fertilizer and in industrial applications for cleaning, refrigeration, and chemical manufacturing.

Hydroxide =
(OH^-)
Contains one oxygen and one hydrogen atom.

Hydroxide is commonly used for cleaning, paper production, water treatment, food processing, and as a component in pharmaceuticals and various industrial processes

Chromate =
(CrO_4^{-2})
Contains one chromium and four oxygen atoms.

Chromate is commonly used for corrosion prevention on metals, as a pigment in paints and dyes, and in leather tanning. It also finds applications in cement and mortar, and as a corrosion inhibitor in cooling water systems.

Sulfate =
(SO_4^{2-})
Contains one sulfur and four oxygen atoms.

Sulfates are found in detergents, shampoos, and other cleaning products as surfactants, which help create lather and remove dirt and oil. Additionally, sulfates are used in agriculture, medicine, and industrial processes.

Can you guess the most commonly used polyatomic ion?

 +  = ??????
(1 atom)   (3 atoms)

The above chart only shows a few of the polyatomic ions formed by those elements. There is no known fixed finite number of polyatomic ions but some other important ones are:

Carbonate (CO_3^{2-}): Crucial in construction, medicine, agriculture, and food production. **Phosphate (PO_4^{3-}):** Most notably used in fertilizers to enhance plant growth, in animal feed supplements, and in cleaning products. **Acetate (CH_3COO^-):** Used in the preparation of metal acetates, used in some printing processes; vinyl acetate, employed in the production of plastics; cellulose acetate, used in making photographic films and textiles.

The most commonly used Polyatomic Ion is *Nitrate (NO_3^-)* : Primarily used in medicine, food preservation, and as fertilizers.

Lithium Ionic Compounds

Lithium itself forms a monatomic cation, Li, which then combines with various anions (including polyatomic ones) to form ionic compounds.

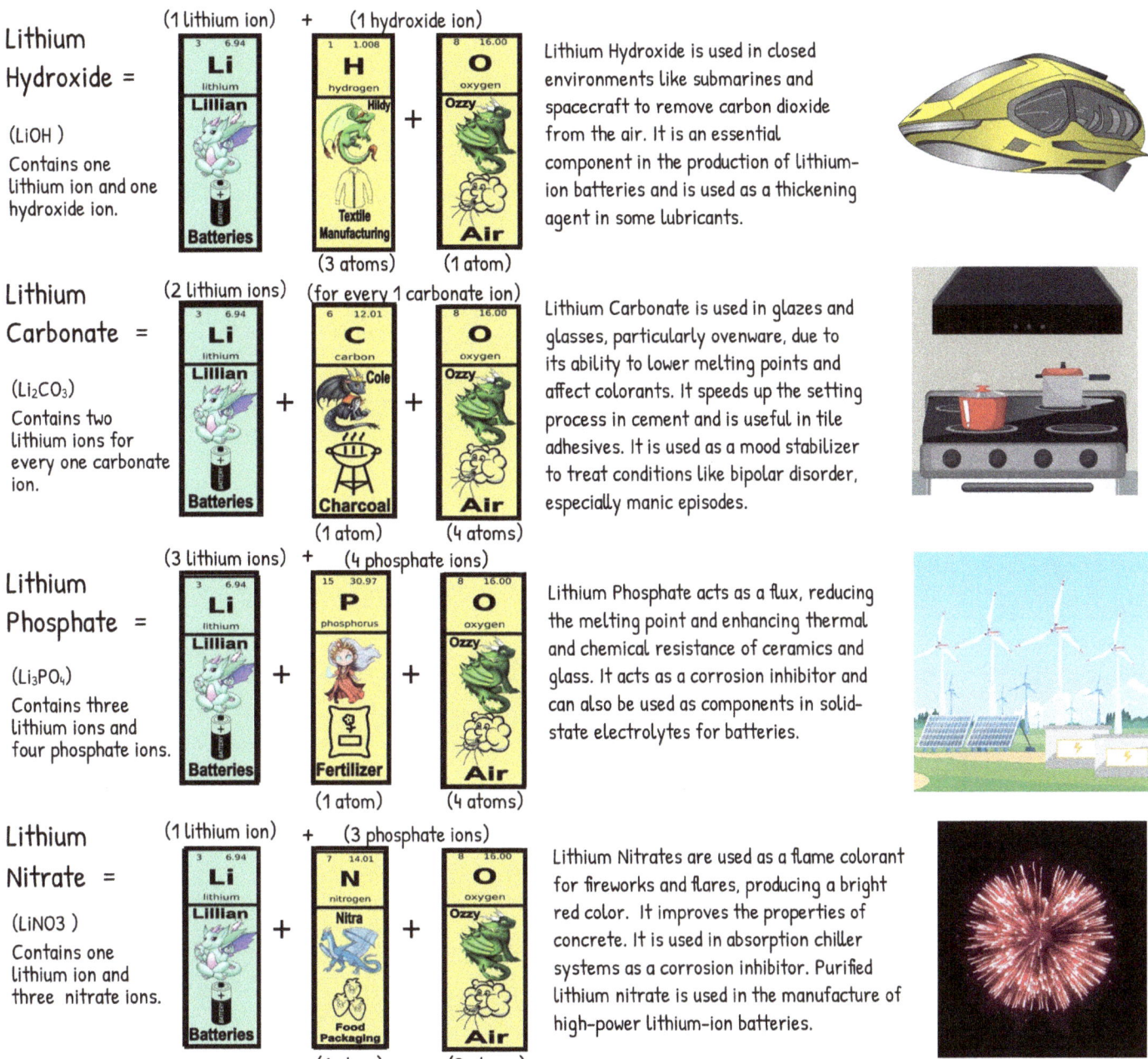

Lithium Hydroxide =

(LiOH)
Contains one lithium ion and one hydroxide ion.

(1 lithium ion) + (1 hydroxide ion)
(3 atoms) (1 atom)

Lithium Hydroxide is used in closed environments like submarines and spacecraft to remove carbon dioxide from the air. It is an essential component in the production of lithium-ion batteries and is used as a thickening agent in some lubricants.

Lithium Carbonate =

(Li_2CO_3)
Contains two lithium ions for every one carbonate ion.

(2 lithium ions) + (for every 1 carbonate ion)
(1 atom) (4 atoms)

Lithium Carbonate is used in glazes and glasses, particularly ovenware, due to its ability to lower melting points and affect colorants. It speeds up the setting process in cement and is useful in tile adhesives. It is used as a mood stabilizer to treat conditions like bipolar disorder, especially manic episodes.

Lithium Phosphate =

(Li_3PO_4)
Contains three lithium ions and four phosphate ions.

(3 lithium ions) + (4 phosphate ions)
(1 atom) (4 atoms)

Lithium Phosphate acts as a flux, reducing the melting point and enhancing thermal and chemical resistance of ceramics and glass. It acts as a corrosion inhibitor and can also be used as components in solid-state electrolytes for batteries.

Lithium Nitrate =

(LiNO3)
Contains one lithium ion and three nitrate ions.

(1 lithium ion) + (3 phosphate ions)
(1 atom) (3 atoms)

Lithium Nitrates are used as a flame colorant for fireworks and flares, producing a bright red color. It improves the properties of concrete. It is used in absorption chiller systems as a corrosion inhibitor. Purified lithium nitrate is used in the manufacture of high-power lithium-ion batteries.

The above chart only shows a few of the lithium ionic compounds formed by those elements. Some other important ones are:

Lithium Acetate ($LiC_2H_3O_2$): This compound contains the lithium ion and the acetate ion, which is a polyatomic ion. It is a buffer in gel electrophoresis for DNA and RNA analysis, a catalyst in various chemical reactions, and a component in lithium-ion batteries.

Lithium Chlorate ($LiClO_3$): This compound includes lithium ions and chlorate ions, which are polyatomic. It is primarily used as a strong oxidizing agent in various applications. It is also utilized in oxygen generation systems and pyrotechnics, particularly for producing intense colors in fireworks

Lithium Permanganate ($LiMnO_4$): This compound contains lithium ions and the polyatomic permanganate ion. Its oxidizing properties make it useful in industrial processes, water purification, and organic synthesis.

Definitions

Atomic Structure of Element: The atomic structure of an element refers to the arrangement of protons and neutrons in the nucleus of the atom, and the electrons in the electron cloud around the nucleus.

Atomic Number: An element's atomic number refers to the number of protons it has in its nucleus. In a neutral atom the number of protons always equals the number of electrons.

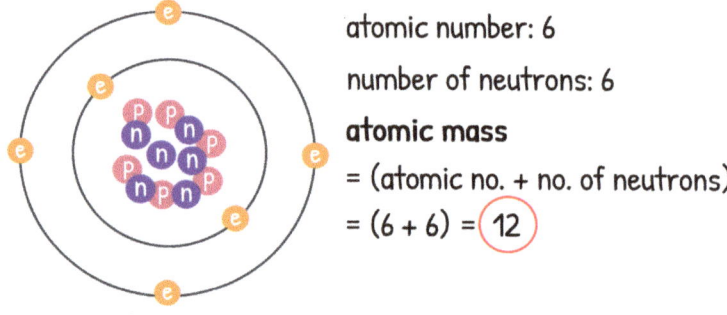

Carbon atom

atomic number: 6
number of neutrons: 6
atomic mass
= (atomic no. + no. of neutrons)
= (6 + 6) = 12

Atomic Weight (Mass) of Element: The atomic mass of an element is how heavy it is. It is made up of protons and neutrons that are in the middle of the element. Some elements have different versions with different amounts of neutrons, but they still have the same amount of protons. The atomic mass is the average weight of all these versions of the element.

Allotrope: Allotropes are different forms of an element that look and act different, but are made of the same stuff. Some elements have more than one form. For instance, carbon can be a shiny diamond or a gray pencil lead called graphite.

Isotope: Isotopes are different types of atoms that have the same parts, like protons and electrons, but they have a different number of neutrons. For example, the three most stable isotopes of hydrogen: protium (A = 1), deuterium (A = 2), and tritium (A = 3).

Crystalline Structure of Element: The crystalline structure of an element is how its atoms, ions, or molecules stick together in a pattern to make a cool crystal shape.

Ferrous and Non-Ferrous Metals: When we say ferrous metal, it means that iron is a big part of the metal. But if there's only a little bit of iron in the metal, we call it non-ferrous. The word "ferrous" comes from Latin and means iron, which is why iron's symbol is Fe.

Ductile Metals: These are capable of being made into long, thin wire or thread. Copper and Silver are ductile metals.

Malleable Metals: These can be hammered or rolled into thin sheets without cracking or breaking. Gold is malleable.

Ferromagnetic: Materials that are strongly attracted to a magnet. Such materials can be permanently magnetized. These include the elements iron, nickel and cobalt and their alloys, some alloys of rare-earth metals, and some naturally occurring minerals such as lodestone.

Magnetostriction—Ther is the term for a special thing that happens to magnetic materials. When these materials get turned into magnets, they also change their shape or size.

Paramagnetic: Slightly attracted to a magnetic field, but do not retain magnetic properties once the field is removed.

Diamagnetic: Slightly repelled by a magnetic field, but do not retain magnetic properties once the field is removed.

Electrical Properties: Conductor—a thing that lets electricity flow through it. Semi-conductor—a special material (usually silicon) that can conduct electricity, but not as well as metal. Insulator (non-conductor)—a material (usually glass) that stops electricity from flowing.

Reactive Gas: These gases are really good at reacting with stuff! They are called "sticky gases" because they can react to things like plastic and wet surfaces when they touch them. These are nitrogen, oxygen, hydrogen, carbon dioxide, fluorine, and chlorine.

Non-Reactive Gas: An inert gas is like a super shy gas that doesn't like to hang out with other chemicals. It doesn't make any new friends by reacting with them, so it doesn't form any chemical compounds. We also call these special gases "noble gases."

What Makes Lithium Seem Magical?

In today's world, the landscape is woven with threads of magic and mystery, where elements can evoke a sense of wonder and enchantment. One such element is lithium, which is a soft, silvery-white metal. This unique substance stands out due to its intriguing properties and has captured the imaginations of many. While modern science has given us a clear understanding of lithium's characteristics, exploring how people perceive this element in a magical light today reveals compelling insights into our ongoing relationship with the mystical.

In our modern world, lithium is often encountered in various mineral forms, such as petalite, which contains lithium and finds its way into different applications, from technology to aesthetics. Consider how everyday lithium-rich materials can spark curiosity and admiration. For instance, lithium-infused products can have a shimmering appearance that reflects exceptional energy and creativity, leading some to view them as gifts of modern innovation. This beautiful quality can easily be compared to magical artifacts in stories, which possess extraordinary attributes that entice and inspire.

The properties of lithium further enhance its allure. In today's realm of modern mysticism, elements that exhibit rare qualities are frequently regarded as enchanting or sacred. Lithium's unique feature of being lightweight and even capable of floating on water carries an ethereal essence, evoking imagery of fairies dancing upon the surface of a tranquil lake. Additionally, lithium's curious reactivity—igniting when it comes into contact with water—can be interpreted as a representation of duality, embodying both lightness and intensity. This duality speaks to modern narratives about balance and the interconnectedness of life, patterns that resonate deeply with themes found in contemporary storytelling.

Exploring how lithium impacts well-being further enhances its mystique. Today, lithium salts are recognized for their calming properties and are often

What Makes Lithium Seem Magical? (Continued)

incorporated into mental health treatments. The practitioners and healers who utilize these compounds may be seen as possessing special knowledge, almost as if they are modern-day sorcerers channeling the essence of lithium to bring healing and tranquility. This perspective allows us to view lithium as a conduit between the ordinary and the extraordinary, amplifying its magical reputation in our collective consciousness.

Cultural narratives also play a significant role in shaping our understanding of lithium. We live in an era where spiritual and scientific inquiries frequently intertwine. The appearance of new materials can spark speculation about their significance, leading to deeper connections with the elements. In many spiritual traditions, elements like earth, fire, water, and air are personified, fostering the belief that elements like lithium can symbolize the presence of benevolent energies or spirits. Thus, lithium becomes more than just a metal; it transforms into a manifestation of elemental forces, earning an enchanting status in our communities.

To appreciate lithium further, we can consider how it was perceived in earlier historical periods, such as the Dark Ages or Middle Ages. During these times, the understanding of elemental properties was often shrouded in mystery. People looked to alchemists and philosophers, those who searched for the philosophers' stone and the secrets of matter. The rare properties of materials like lithium would likely have captured the imagination of these seekers. They might have attributed extraordinary powers to it, seeing the metal as possessing magical qualities that could influence health, longevity, or even enlightenment.

Medieval scholars often believed that certain metals had spiritual significance. The alchemists, in particular, sought the transmutation of base elements into noble ones. The smooth, glimmering surface of lithium could very well have been seen as a sign of its otherworldly power—perhaps even thought to be a gift from the heavens. As miners unearthed these minerals, tales would have developed around them, creating folklore that weaves together the natural world's beauty and a mystical perspective rooted in the unknown.

As we navigate our modern scientific world characterized by observation and inquiry, lithium's rarity and fascinating reactivity can inspire stories of innovation and exploration. Today's storytellers weave narratives about individuals who harness the powers of such elements, embedding lithium into the fabric of contemporary folklore.

Meet Lillian, The Dragon With The Lithium Tipped Tail

Lillian is a lovely green dragon who calls an enchanted valley her home. This magical place is filled with bright, colorful flowers that sway gently in the breeze and sparkling streams that sing sweet songs as they flow. Lillian's smooth green skin glistens like shiny satin when the sunlight hits it, making her look even more beautiful. Her large, magnificent wings are a stunning purple, shimmering with shades of lavender and violet that catch everyone's attention. Lillian is not just an ordinary dragon; she possesses a magical ability to wield the power of Lithium at the tip of her extraordinary tail. With a gentle caress from her tail, Lillian can soothe the hearts of those in turmoil. As she trails her tail across the chaotic souls around her, a wave of calm spreads like soft ripples on a still pond, silencing their fears and anxieties.

Her powers extend beyond mere tranquility; she can absorb the tumultuous emotions that surround her and transform them into brilliant bursts of energy. This light can be released in various forms—a dazzling display of sparks that fill the air with laughter, or radiant beams that ignite creativity in the hearts of artists and dreamers. Lillian takes pride in her unique ability, always eager to help those in need and bring joy to her valley.

One sunny afternoon, Lillian had just finished a joyous flight over the valley when she heard a cacophony of shouts near the old wishing tree. This ancient tree was a sacred place where villagers often gathered to share their hopes and fears. Lillian felt a surge of concern. With her heart racing, she swooped down to see what was amiss. As she landed gracefully, she saw a group of children huddled together, their faces twisted in worry.

"What's wrong?" she asked gently, her voice soft and soothing like a warm breeze.

The children looked up at her with wide eyes, fear written across their small faces. "The tree!" one child whimpered. "It stopped listening. We wished for fun and adventure, but nothing happened!"

Lillian curled her tail around the children, hoping to soothe their worry. "Shhh," she whispered softly, invoking her Lithium power. A warm glow enveloped her tail, and soon, the children were lulled into a state of calm. Their chaotic fears began to drain away, replaced by a comforting sense of serenity.

With her heart swelling with empathy, Lillian thought for a moment. "What if I showed you the magic of believing?" The children seemed intrigued, their curiosity piqued.

She lifted her wings high and, channeling her Lithium energy, she unleashed a cascade of shimmering sparks into the air. The sparks twinkled and danced in the sunlight, filling the space with a warm glow. They fluttered like tiny fairies around the children, illuminating their faces with wonder.

One by one, the children began to giggle and clap their hands. Lillian watched as their imaginations ignited, their minds filling with visions of fantastical adventures. "Look!" shouted a girl with curly hair, pointing to the sky. The sparkles had twisted into the shapes of dragons, castles, and magic potions. The valley around them seemed to come alive, bursting with color and excitement.

Seeing their joy rekindled something within Lillian. She felt a stirring in her heart, a glimmer of hope. Could it be possible to use her power to restore the old wishing tree? The tree had stood in the village for generations, known for granting wishes to those with pure hearts. However, it had grown tired and faded over the years. Now, it seemed almost forgotten. With a determined heart, Lillian approached the gnarled trunk of the tree. It was old and weary, but deep down, Lillian could feel the magic still pulsing within it, like a heartbeat waiting to come alive again.

She caressed the rough bark with her tail, feeling a connection to the tree she had known all her life. Summoning her energy, she focused deeply. For a moment, the tree trembled, responding to her call. Lillian poured her Lithium power into it, creating a soft, glowing light that enveloped the trunk. To her delight, she watched as colorful flowers began to bloom at its base, their petals stretching out as if they were waking from a long sleep. A gentle breeze rustled through its leaves, whispering promises of wishes yet to come.

As she concentrated, Lillian realized that the wishes the villagers had made over the years were not lost; they had merely been waiting for the right spark of encouragement. It was then that she felt the magic of the tree returning, stronger than before. With her help, the ancient tree pulsed with renewed vigor, its branches swaying eagerly as if ready to listen once more to the hopes and dreams of the villagers. Lillian smiled, filled with joy, knowing that together, they could make wishes come true again.

Before long, the villagers returned, their amazement evident as they saw the tree thriving again. Bright flowers bloomed all around it, and the sight made their hearts soar. Lillian beamed, knowing that her Lithium power had reignited the hope of an entire community.

As the sun began to set, casting a golden hue over the valley, Lillian spread her wings wide. With great joy, she soared into the twilight sky, leaving behind a shimmering trail of light that promised creativity, adventure, and the assurance that chaos could always yield to calm.

From that day forward, Lillian was not just another dragon; she became a guardian of dreams, weaving tranquility amidst the turmoil of life. She reminded all who crossed her path that sometimes, a little magic and kindness are all it takes to turn fear into wonder. The children would often find her, and together they would create new wishes beneath the old tree, knowing that with a little belief, anything was possible.

Enjoy This Coloring Page Featuring

Lillian The Wizard With The Lithium Tipped Tail

Sample Page From Magical Elements of the Periodic Table Presented By The Elemental Dragons

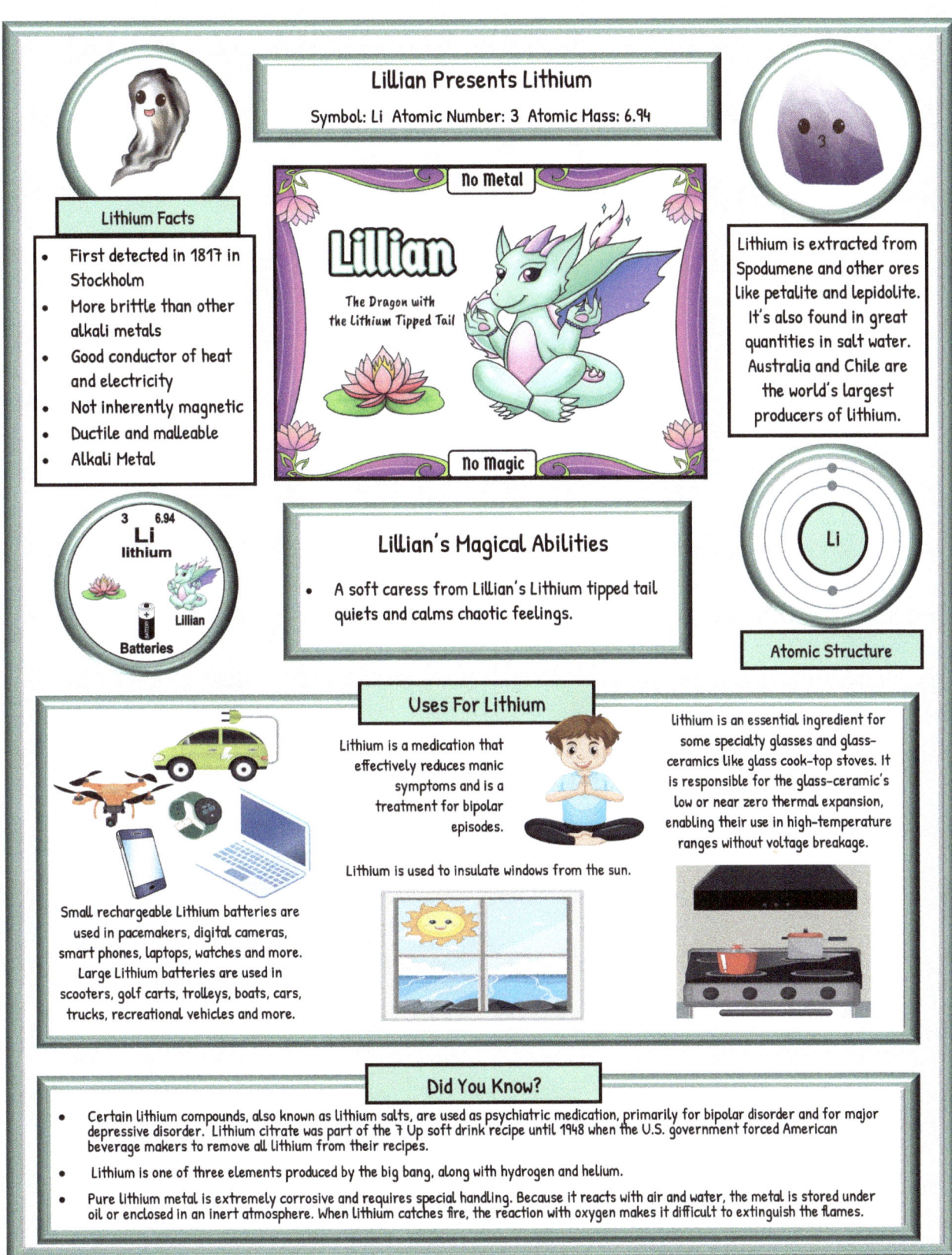

Magical Elements of The Periodic Table

Create Your Own Magical Dragon Elemental

Lillian — The Dragon With The Lithium Tipped Tail

Symbol: Li   Atomic Number: 3   Atomic Mass: 6.94

- Magical Clan Crest Symbol  
- Alkali Metal
- Atomic Structure (Li)
- Lithium is extracted from Spodumene
- Lithium Periodic Symbol

Lillian's Magical Abilities

- A soft caress from Lillian's Lithium tipped tail quiets and calms chaotic feelings.

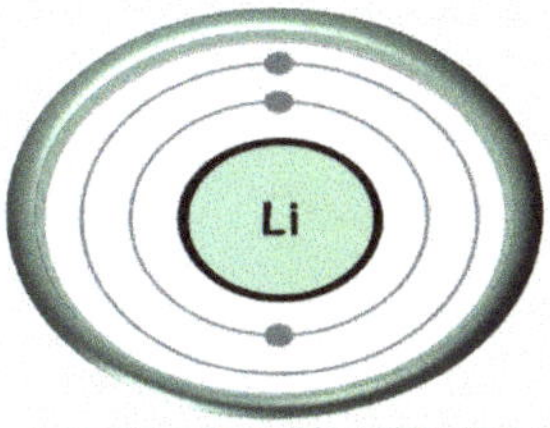

Magical Elements of The Periodic Table

Students may either use a program like power point to cut and paste clip art into a Magical Dragon Elemental Blank or, if they wish, they may draw everything themselves.

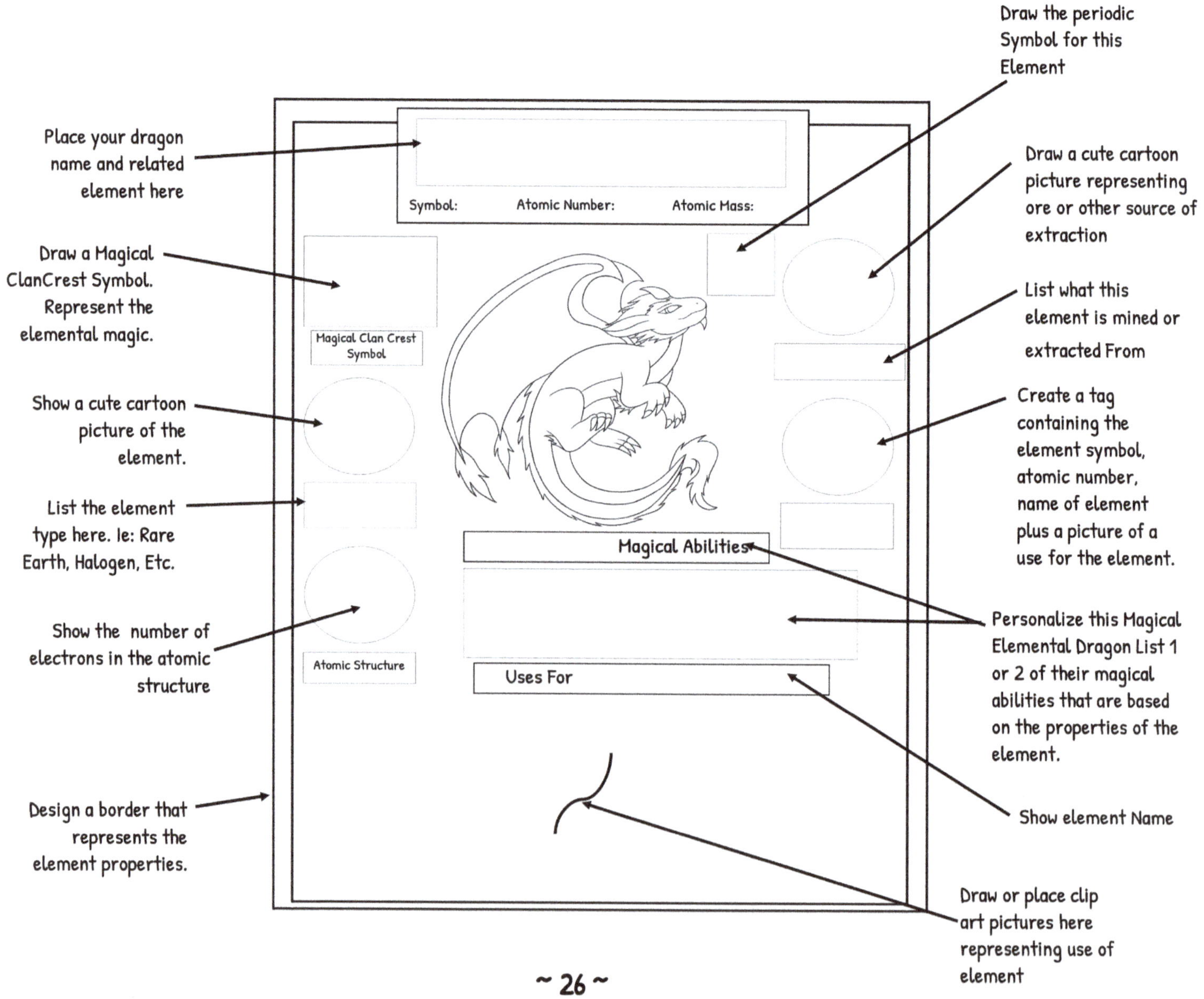

Labels on the template (clockwise from top):

- Draw the periodic Symbol for this Element
- Draw a cute cartoon picture representing ore or other source of extraction
- List what this element is mined or extracted From
- Create a tag containing the element symbol, atomic number, name of element plus a picture of a use for the element.
- Personalize this Magical Elemental Dragon List 1 or 2 of their magical abilities that are based on the properties of the element.
- Show element Name
- Draw or place clip art pictures here representing use of element
- Design a border that represents the element properties.
- Show the number of electrons in the atomic structure
- List the element type here. Ie: Rare Earth, Halogen, Etc.
- Show a cute cartoon picture of the element.
- Draw a Magical ClanCrest Symbol. Represent the elemental magic.
- Place your dragon name and related element here

Template fields: Symbol: Atomic Number: Atomic Mass: Magical Clan Crest Symbol Atomic Structure Magical Abilities Uses For

~ 26 ~

Symbol: Atomic Number: Atomic Mass:

Magical Clan Crest Symbol

Atomic Structure

Magical Abilities

Uses For

Magical Dragon Elemental Research Sheet

Before starting your Magical Dragon Elemental graphics page, do some research on your chosen element.

Name of Magical Dragon:	
Dragon's Magic Power Based on the Element's Properties:	
Magical Clan Crest Symbol:	
Element Name:	
Element Symbol:	
Atomic Number:	
Atomic Mass:	
What year and where was this Element discovered?	
Who discovered this Element?	
Element Group:	
Element Period:	
Element Family Name:	
State of Element At Room Temperature:	
What is Element Mined or Extracted From?	
Is Element Magnetic?	
Does Element Conduct Electricity?	
Where is the Element commonly found in Nature?	
What is 1 alloy of the Element? How used?	
What is 1 compound of the Element? How used?	
Name the most common use for this Element:	
Name a little known use for this Element:	
Name one more use for this Element:	
Interesting and Fun Facts:	

Magical Unicorn Elemental Research Sheet

Before starting your Magical Unicorn Elemental graphics page, do some research on your chosen element.

Name of Magical Unicorn:	Ghel The Gold Horn Unicorn
Unicorn's Magic Power Based on the Element's Properties:	Ghel can see past, present and future. She is empathic and can sympathize with the feelings of other. They say she has a heart of gold.
Magical Herd Crest Symbol:	An open heart with a Celtic Trinity Knot.
Element Name:	Gold
Element Symbol:	Au— Comes from Aurum which is the Latin word for Gold.
Atomic Number:	79
Atomic Mass:	196.97
What year and where was this Element discovered?	Around 4,600 BCE in Bulgaria
Who discovered this Element?	Unknown
Element Group:	11 on Periodic TAble
Element Period:	6 on Periodic TAble
Element Family Name:	Gold is a Noble Transition Metal
State of Element At Room Temperature:	Solid
What is Element Mined or Extracted From?	Quartz Veins. It is also found in gravel in streams.
Is Element Magnetic?	It is Diamagnetic. It's only weakly magnetized when placed in a magnetic field.
Does Element Conduct Electricity?	Gold is a great electrical conductor used in printed circuitry of computers.
Where is the Element commonly found in Nature?	One of the largest deposits is found in the United States in Arkansas.
What is 1 alloy of the Element? How used?	White gold is an alloy of gold, palladium, nickel and zinc.
What is 1 compound of the Element? How used?	Gold Phosphide is a semiconductor used in high power, high frequency applications and in laser diodes.
Name the most common use for this Element:	Jewelry
Name a little known use for this Element:	Acupuncture needles
Name one more use for this Element:	Gold is used in airbags in cars.
Interesting and Fun Facts:	Gold was used in ancient Egypt to fill decayed teeth. Gold thread is incorporated in astronaut spacesuits to protect them from the heat of the sun.

Write a paragraph below to describe your magical dragon elemental. Based on the information obtained from research of your chosen element, how did you determine your dragon's name? What are your dragon's magic powers? What are their likes/dislikes, strengths/weaknesses, personality traits? What color is your dragon and why did you pick that color? What is your dragon's Clan Crest Symbol?

Magical Unicorn Elemental Sample Description

Ghel The Gold-Horned Unicorn

This magical unicorn has a golden horn and hooves that glow like the sun. Her hide is honey-gold and her flowing mane and tail are golden-blonde.

Ghel is a member of the Metal Horn Unicorn Tribe from Unimaise. Gold is linked to the heart chakra because it holds a warm energy that brings soothing vibrations to the body to aid in the healing process. Ghel's Magical Herd Crest symbol is an open heart with a Celtic Trinity knot which indicates that she can see past, present and future. She is empathic and can sympathize with the feelings of others.

Being empathic does not make Ghel a weakling. She is brave with strong opinions and is a true champion to those she loves. She is known as the unicorn with the "heart of gold". When she places her horn on the heart of another, she senses their future.

The name Ghel is an Indo-European word which means yellow. The word "gold" most likely has its origins in the word "Ghel".

Gold is known to possess spiritual powers that bring happiness, peace, stability and luck to those who wear it. Scientists say that all the gold in the world comes from the collision of neutron stars.

Though most other magical unicorn elementals get along well with the gold horn unicorn herd; Ghel, and others like her, must be very careful around the Quick Silver Herd - as gold dissolves in mercury.

Do Your Middle Graders Want To Know More From The Magical Elementals About The Periodic Table?

Get the accompanying books in print at all online book stores. Get the books and accompanying activities at MagicalPTElements.

Available Now

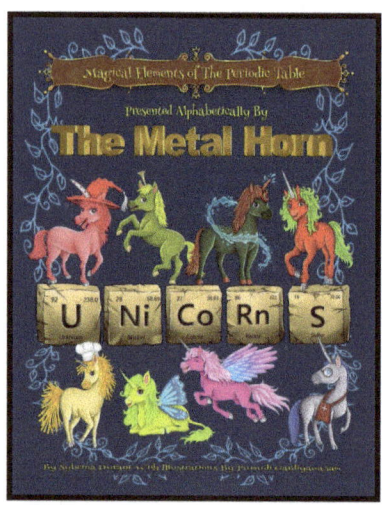

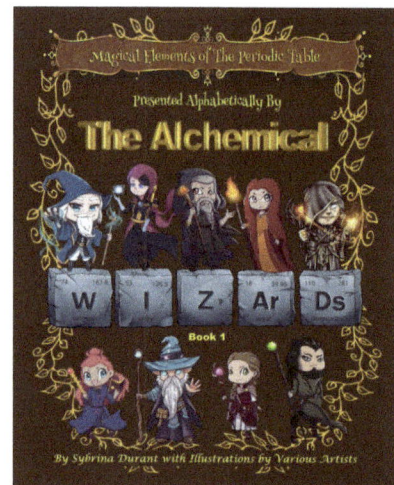
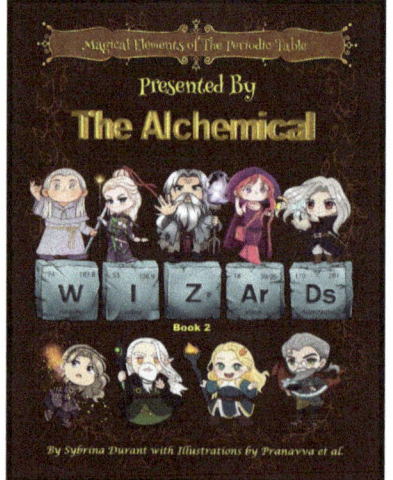
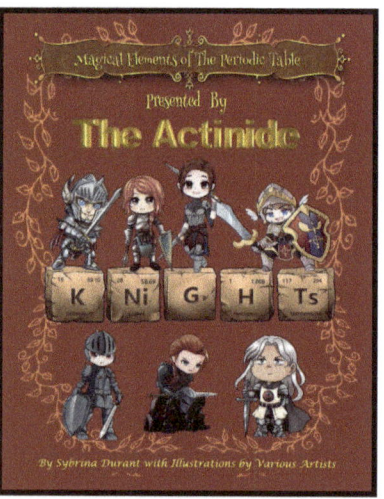

Coming Soon

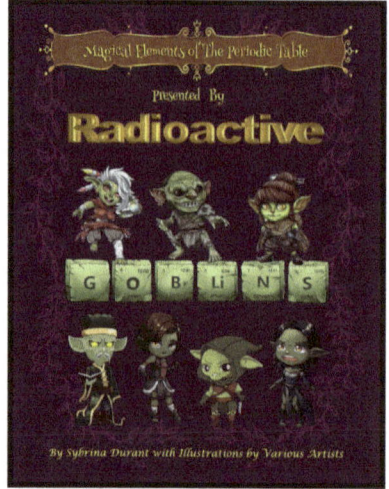

Get These Trading Cards Sets Featuring The Magical Elementals Representing The Periodic Table Elements

at https://bit.ly/4OoEUBr

Unicorns, Dragons, Wizards or Knights?

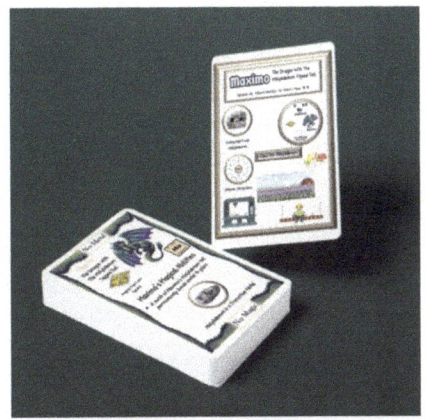

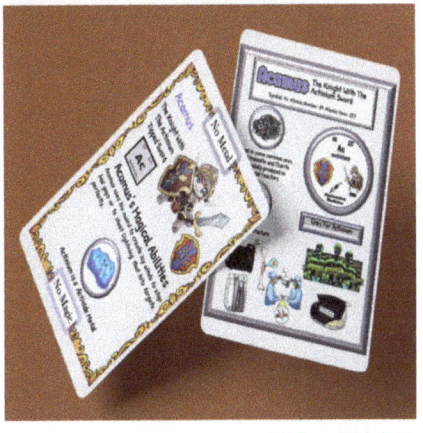

Collect Them All

These magical elementals are ready to help make learning the periodic table more fun. Get all books and related activities today.

Learn More About all of the Periodic Table Elementals. Get all of the "No Metal No Magic" Books Featuring Individual Elements at MagicalPTElements.com

Also Available From Sybrina Publishing
Magical Elemental-Themed Periodic Table

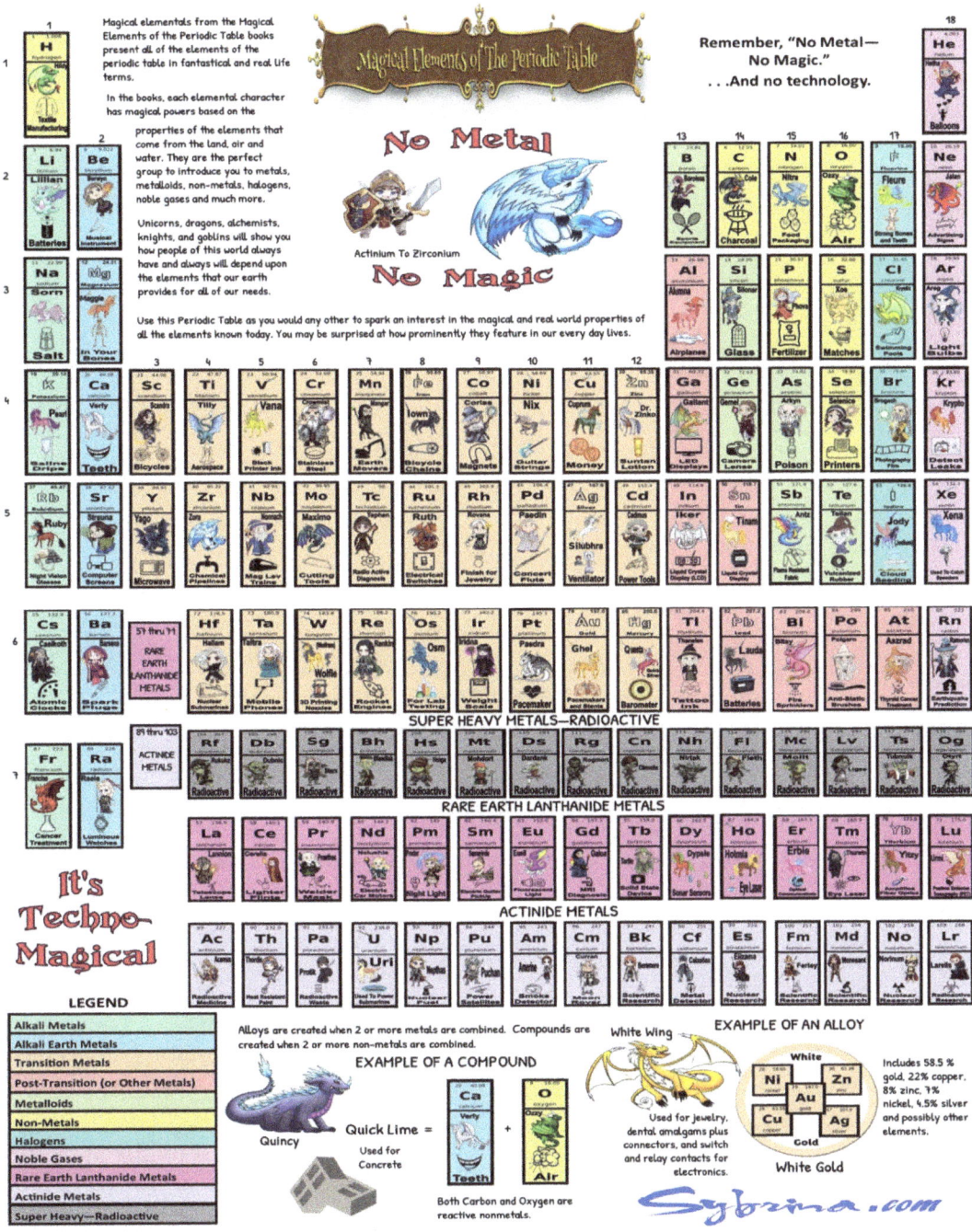

Would you like a 24" x 36" poster of the Magical Elemental-Themed Periodic Table from The Magical Elements of the Periodic Table books? The best place to get a high quality poster size print is at https://bit.ly/49QMxBT   It will look great on a classroom or kid's room wall.

Get These Fun Elemental Periodic Table Activities at

MagicalPTElements.

Unicorn Periodic Table Bingo—Comes with 32 unique Bingo cards. Magical Elementals Bingo comes with 36.

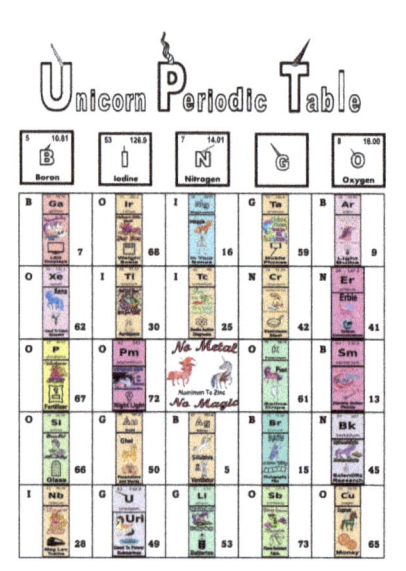

Magical Elemental Game Cards—Makes great prizes. Fun to trade, too.

plus Unicorn Horn Alphabet

CLIP ART FOR YOUR GRAPHICS

Also browse activities at
https://www.magicalPTelements.com
for all kinds of printable downloads to make learning fun.

Printable Magical Elemental Activity Downloads

Fun Way For Students To Learn The Elements Of The Periodic Table

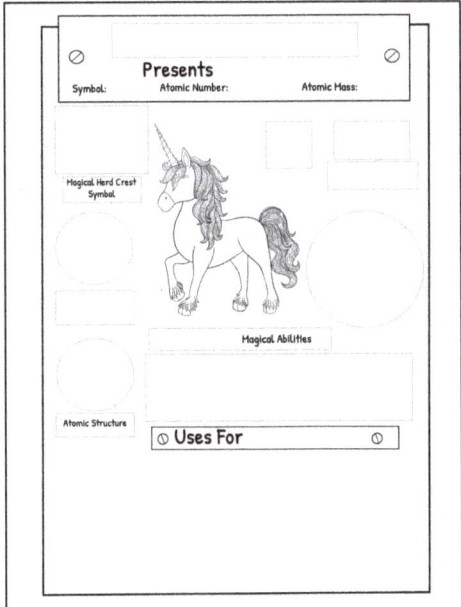
Blank Unicorn Element Card

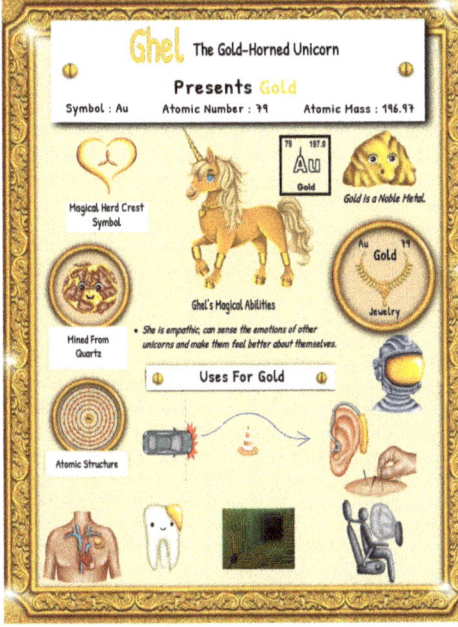
Sample Unicorn Element Card

Blank Research Sheet

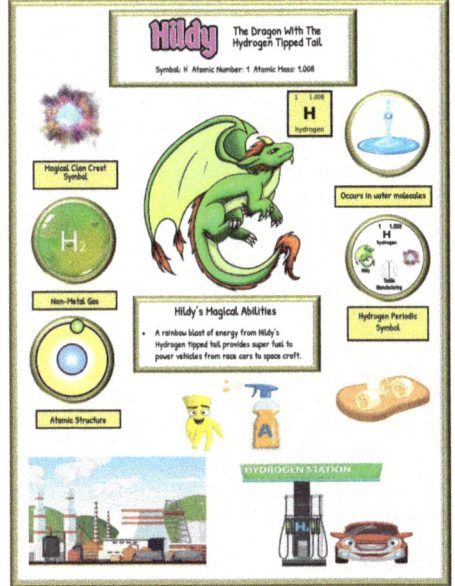
Sample Dragon Element Card

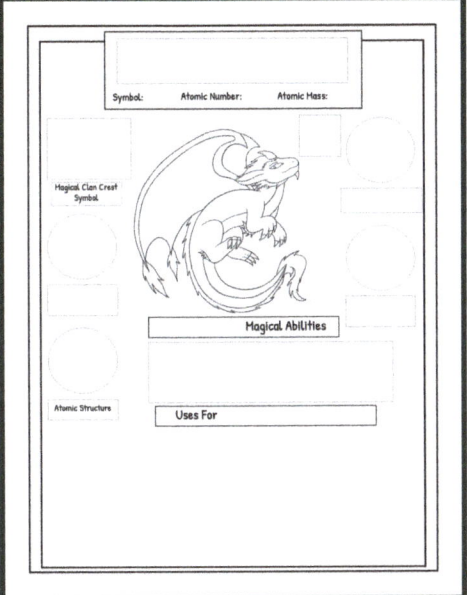
Blank Dragon Element Card

Blank Research Sheet

Using the sample Magical Elemental cards provided, have students select an element from the Periodic Table and a Magical Elemental Card Blank to create their own Magical Elemental Card. The blank and sample cards do not have to match.

You will receive a pdf containing either 26 unicorn or 26 dragon sample cards and blanks to be printed on 8 1/2 x 11 sized paper or card stock. The pdf also contains a Magical Elemental Research Sheet for the students to work on before creating their unique Periodic Table Elemental. They will also write a short paragraph describing their Unicorn or Dragon Elemental from that research.

Get These Fun Elemental Periodic Table Activity Sheets at *MagicalPTElements.com*

Don't Forget To Get A Tee Shirt

Featuring Your Favorite of the

118 Elements from the Periodic Table

Available in adult and kid sizes in many colors.

https://amzn.to/47NVZWN

This is Lillian's Tee Shirt Graphic.

Get it at https://www.amazon.com/dp/B0D3Y7PQ72

Dear Reader

I hope "No Metal No Magic Element 3 — Lithium, Presented By Lillian, from The Magical Elements of the Periodic Table Book Series" with illustrations by Pranavva et al, has helped you learn some fun and interesting things about the magic of the element, Lithium.

This is one of what will eventually be 118 books featuring periodic table elements presented by unicorns, dragons, wizards, knights and goblins. Keep checking regularly. Every one of the elements are amazing and very necessary to our everyday lives. All of the elements are

Techno-magical.

A lot of research went into every page of this book as well as the Magical Elements of the Periodic Table Books. There are just too many references to publish in this book but you can read and research them all at MagicalPTElements.com/MAUPT or /MDAPT or /MW1PT or /MW2PT or /MAKPT. There, you can also access book related activity sheets and games to help make the learning process more fun.

Get ready made trading cards, lapel pins, tee shirts and more based on this book from Sybrina Publishing's No Metal No Magic Collection at Zazzle - **http://bit.ly/3km64Wg**

Would you like a 24" x 36" poster of the Elemental-Themed Periodic Table in this book? The best place to get it is at **https://bit.ly/49QMxBT** They have the sharpest images of any other poster printer around.

The Magical Elements of the Periodic Table books came into existence because of my Blue Unicorn—Journey To Osm books. If it weren't for their magical powers, based on the properties of the metals of their horns and hooves, I would have never come up with the idea to relate magical creatures to the periodic table. There's a metal horn unicorn story for every age group and they are all available at MagicalPTElements.com

If you enjoyed this book
please leave a nice review
at your favorite online book site.

No Metal No Magic
Song Lyrics

No metal, no Magic

No metal, no Magic

I can think of nothing more tragic

Than to have no metal or no magic

Metal makes everything magical.

Just ask a unicorn. . .

Preferably, one with a metal horn.

They'd say No metal, No magic.

Metal makes everything techno magical.

No metal, No magic

for two-leggers or unicorns.

No metal, No magic

Metal makes everything techno magical.

No metal, No magic

It's techno magical.

No metal, No magic

It might be very hard to believe but with

No metal, No magic

There'd be no technology.

No metal, No magic

Listen to this song at https://youtu.be/tcB8KDWAd8w

Watch the book trailer at https://youtu.be/NIX9fE7GJRI

Blue Unicorn

Ebooks, Audio and Print Books Available at all online book stores.

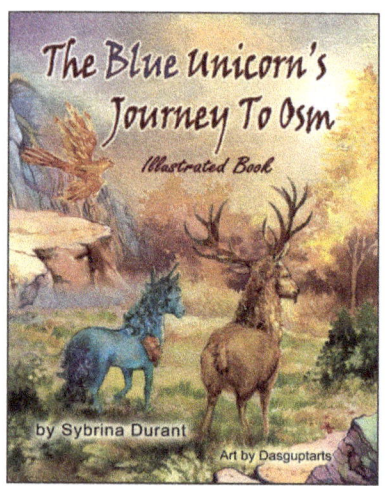

Illustrated Book

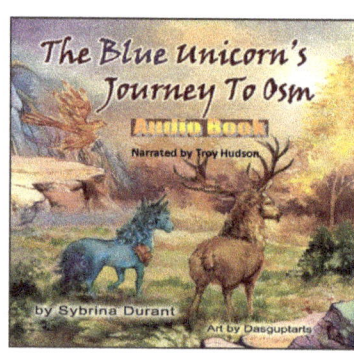

Audio Book

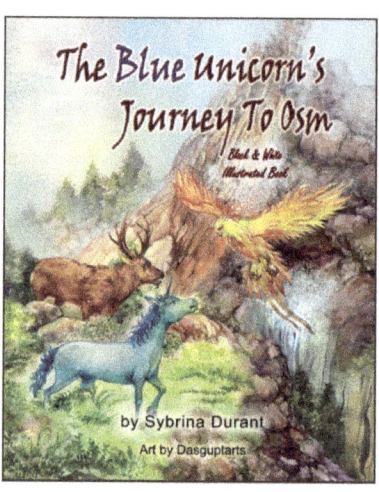

'Read & Color' Book For Teens

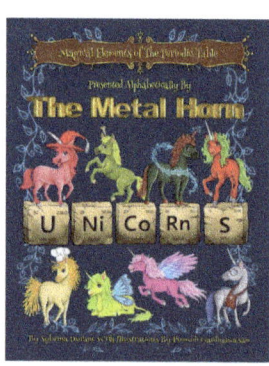

Unicorn Periodic Table Book

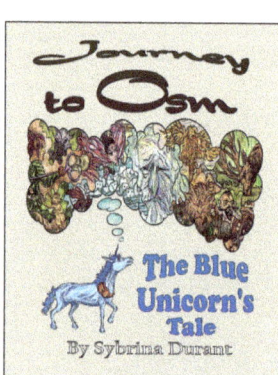

Fantasy Novel

Coloring Book & Character Introduction

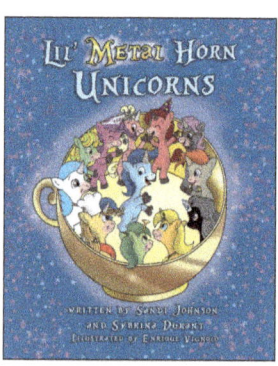

Picture Book For Kids

Get these and more at
MagicalPTElements.com
and Sybrina.com

Excerpt from Chapter 24 (Rag-Bag Magic) of the novel,

Journey To Osm—The Blue Unicorn's Tale

Back then, most places throughout MarBryn had wizards and sorcerers of some ability, or another. Most were trained in the ways of magical arts by the unicorns as part of their outreach program. Some two-leggers developed practical magical skills like making delicious feasts of tasty food appear out of thin air or purifying murky water around the land.

Others went through more extensive training to learn battle magic—like shooting powerful streams of energy from their swords.

Some were taught the art of holding the glow of the sun in magical globes, bringing light into the dark of night. These magical lights warded off the evil beings that were new and frightening products of dark magic.

With the rise of the sorcerer Magh, magical defense arts had become more important. The highest level of magical training involved sensing when others were in danger and learning to see into the future. Very few two-leggers ever reached that level because magic wasn't inherent in them the way it was for the unicorns. The metal of their horns and hooves were part of them as well as the very makeup of their blood, but two-leggers relied on learned magic via potions, charms, and incantations that required help from ingredients and forces more mystical in nature than any two-legger was ever born to be. Of course, controlling magic and projecting your intentions went far beyond merely following a recipe of sorts. It took being in touch with nature and the various elements to get the response a wizard desired. Much trial and error went into it, as well as faith and trust and the motives of the spell caster.

Magic was and is a practice that is never quite perfected even for the unicorns who must continue to hone and learn how to harness their powers. On rare occasion a wizard and unicorn had formed enough trust and a steadfast bond that prompted the unicorn to gift the wizard with a wand or staff embedded with the smallest sliver of metal from one of their hooves. This was rare but had happened and of course so had the desire for more power. Magh wasn't the first sorcerer with lust for more and throughout history there had been a handful of heinous acts against unicorns from those seeking their magic. Prior to Magh those wizards had failed to circumvent the protections nature had infused unicorn magic with so even after harvesting metal from their horns or hooves these sorcerers had gone mad trying to bend the will of nature and actually use their ill-gotten gains.

Magh, too, had gone mad or perhaps he'd already been so, but somehow he'd managed to harness the metal he harvested from his victims and continued to grow stronger rather than completely lose his mind like the others. How exactly remained a mystery to the unicorns and everyone else.

The wizards who'd honed their craft with the help and blessing of the unicorns tried to solve the riddle and even the best of them failed to uncover that secret. Some of MarBryn's natives took to magic naturally, while others struggled with the concept but no matter how adept they were. All of them fought valiantly against Magh's magic because with even a shred of knowledge they understood how the power shift would ultimately play out. They fought to the end but, in the end, only one sorcerer remained in MarBryn and now, Magh was in total control. But still, he was not satisfied. He wanted to control every living creature in the land.

Get The Novel at MagicalPTElements.com

www.ingramcontent.com/pod-product-compliance
Lightning Source LLC
Chambersburg PA
CBHW040004080526
44586CB00027B/2879